U.F.O. Contact/Abduction: A Healer's Perspective

A Handbook for the Contactee, Abductee and His/Her Family

trisha keen, a./c.h.t.

ISBN 0-9632153-0-2

To Contact the Author, Healers, or Cover Artist mentioned in this book please write to:

Trisha Keen, A./C.H.T.
c/o A.T.R.I. Publications
2040 Polk Street, Suite 212
San Francisco, California 94109

Cover Art By:
Catleen Clark

1st. Printing, October, 1991
2nd Printing, February, 1992

Printing By:

Atlas Press
1350 Bush Street
San Francisco, CA. 94109
(415)- 474-4061

Printed in the U.S.A.

Special Thanks To:

Joni and Eric Haynes
Jerry Whitlock and Mark Bankins

Dedication

This book is dedicated to all who have the courage to seek the I AM Principle in themselves and others.

To **Ms. Eva Fried** and **Ms. Diana Freitas,** two of the most remarkable women I have ever met. May their courage, determination and unconditionally loving Light shine forth to all who have lived with illness, fear, ridicule and pain.

To my clients who have walked the path from victimization to freedom, self-empowerment and healing with courage and love in their hearts. There is no way they will ever know the love, respect and admiration I have for them!

Acknowledgments

First and foremost, my loving appreciation goes to my son, Richard, and Ms. Jennifer Buckingham. Avid skeptics who have supported my efforts, driven me crazy and kept laughter and loving present when I tended to take life much too seriously.

U.F.O. Researchers, Dr. Jacques Vallee, Tom Dongo, Don Christian, Ian Christopher, Jack Malin and Ted Oliphant. Men who, knowingly or unknowingly, walk a path of Light that shines brightly.

Joni Haynes, Arline Bloom, Martha Ellis, Ray Langevin and Michael Ohayon, for their typing, computer time, editing and loving support of this project. Last but by no means least ... my loving and patient Guides who have come forward to teach me, and who aid in the healing processes of the contactees and abductees who walk through my door.

Note To Readers

A.T.R.I. plans subsequent books delving more deeply into the hypotheses and material in this book. If you have questions or comments regarding the material presented please write us. Questions will be noted for use in forthcoming publications.

If you are a Contactee or Abductee in need of help, or simply in need of communicating with someone who understands what is happening with you please drop us a note. Be sure to include your address and telephone number. Someone from our contactees and abductees group will contact you either by telephone or letter, whichever is most feasible, as soon as possible.

If you are interested in **Career Certification Programs** in Alchemical Hypnotherapy or in simply learning the process as an aid in Personal Transformation, please write for the 1992 Training Schedule.

Workshops, Seminars and 'Joy-Shops' are available in many areas of interest including Intuition Development for the Medical and Psychiatric Communities. A Speakers Bureau is available for the education of groups or organizations interested in learning about U.F.O'S., Contact and/or Abduction.

Please Address All Correspondence To:

Trisha Keen, A./C.H.T.
c/o A.T.R.I. Publications
2040 Polk Street
Post Office Box 212
San Francisco, CA. 94109

Table of Contents

Glossary of Terms

Akashic Records: A group memory that holds information from all lifetimes; past, present and future.

Alien: A being from another planet or Universe that has no concept of love or compassion.

Aura: The energy field that surrounds and emanates from all living things; plants, animals and humans alike.

Body-Being Connection: A cord attached to a chakra so the soul may leave the body.

Conscious Out of Body Travel: A process of the soul partially leaving the body to travel through the Universe or other Dimensions while one is in a waking state.

Cellular Memory: Memory held in the physical body. This memory may be carried from lifetime to lifetime.

Dimension: Webster's New World Dictionary defines Dimension as: "n. [to measure] 1. any measurable extent, as length, width, etc..2. [pl.] measurements in length, width, and often depth. 3. extent; scope." In more metaphysical terms Dimension refers to a place in the space / time continuum rather than a vibrational rate or density factor.

Discarnate Entity: A soul no longer connected with the physical body.

Extraterresterial (E.T.): A being from another planet, universe or dimension that functions from a position of love and compassion.

Keeper of the Akashic Records: The being one calls upon to gain access to his or her personal records from the group memory of the universe.

Universal Law: Laws governing the universe

Whacks: A hole or break in the auric field surrounding the body caused by a negative thought form or insult.

A Can of Worms

Have you ever been raped? Did you have an abusive childhood? Well, maybe not ... BUT ...

A friend *of yours has, in a moment of need, shared his or her experience of rape or abuse with you...*

Do you shut him or her out of your life because rape or abuse is **not your reality?**

If a friend shares an experience regarding U.F.O. sightings, contact or abduction is he or she a liar, or crazy because the experience is **not your reality?**

The existence of alien space craft, alien contact and alien abduction has been researched and documented for centuries.

Since the 1940's alone there have been numerous stories, and now media release of information, about U.F.O. crashes and sightings; government cover-up of information; alien contacts and abductions worldwide, yet many world leaders and citizens alike still refer to this experience as a phenomenon or a fantasy.

I feel the proof is here. It comes from people all over the world who share experiences. People who wake up one day with a small, detached memory or picture of some dream scenario that leaves them with a feeling, a knowingness, that something has happened in their lives that they must uncover and possibly stop, or at the very least find out about.

Isn't it about time society left the denial stage and allowed the contactee or abductee, the person researched, the help he or she needs to heal, without fear of government reprisal or public ridicule?!

In the next few pages I will share a brief story about how I became involved in the U.F.O. area; some hypotheses regarding contact and abduction; examples of some average scenarios and exercises that may aid the person experiencing contact or abduction, along with some interesting theories pertaining to physical health and abductees.

So...On With a Brief Personal Story

The following conversation took place late in the evening on Sunday, February 5th, 1989. Since that time I have had over one hundred and thirty similar calls and my life has been one mind stretching experience after another!

"Hello?" I answered to the shrill ring of the telephone.

There was a long pause on the other end of the line so I repeated my greeting.

"Hello?, this is Trisha speaking, may I help you?"

"I hope so, I was told you could help me," replied a soft, fear filled voice. "I have been abducted and don't know what to do about it."

"Abducted? By whom? When?" I asked, as my mind began racing.

"I was abducted by aliens ... you know, a U.F.O. and, and I need some help coming to terms with it. My life has been a real mess. No one believes me, I'm not crazy! Will you help me?"

I could hear panic rising in the caller's voice.

"Who suggested you call me?"

"I don't know but, I did write down your number, will you help me?"

My immediate reaction to this call was simple ... it had to be a prank! The next day I received three more calls with similar stories. These calls continued for several months.

After this initial call I went in search of people I thought were experts in the field of U.F.O.'s.

The people I was guided to were researchers. Unfortunately, I never met nor was referred to any therapists during this search.

One of the researchers I met was convinced I was an abductee and asked me to do a session with him. What I learned during this session regarding the treatment of abductees or contactees is why I finally decided to see clients who had had the abduction or contact experience.

First of all, during my session, we couldn't locate any contact or abduction at all in this life. At one point I slipped back, prior to birth; I remember, I was standing before some sort of council; a being was walking toward me.

This being didn't resemble the researchers' preconceived idea of what it should look like. What I saw resembled a human-like form.

The researcher didn't want to hear what the being was saying and kept badgering me, insisting I see something that was not there.

During the next few weeks I felt a great deal of confusion over this session. The information being shared by this being was very loving and positive.

The being was explaining ways for researchers

to help people who sought them out. It kept repeating: "Not all contact is negative; it can be a very positive and spiritual experience. We are here to help if you will only ask."

I couldn't remember everything that occurred but the gentleness of this entity stayed with me. Also, the idea of contact or abduction being a spiritual experience fascinated me.

There are many psychiatrists and therapists who are willing to help those who feel they have had the abduction experience by facilitating memory recovery. The standard methodology used is hypnosis.

Webster's New World Dictionary defines hypnosis as: a sleeplike condition physically induced, in which the subject responds to the suggestions of the hypnotist.

Okay, let me backtrack for just a second ... the night I received the first phone call was the same night that I returned from the Alchemical Hypnotherapy Institute. I had just that day received my certification in the Alchemical Process.

Alchemical Hypnotherapy is a spiritual science of transmutation. Its name is borrowed from the ancient alchemist who had as their purpose transforming the weight of negative emotions into the gold of enlightened awareness.

It goes beyond symptom removal to personal integration and spiritual awareness. It is interactive, and the suggestions which are used come from the client's inner guidance, not the therapist. This insures that the paths taken are appropriate for each individual.

I took this training because I believe it to be the most empowering and healing process a person could ever experience; not just emotionally but physically as well.

My answer was staring me in the face. I had been given a gift ... a process that would allow an abductee or contactee to empower themselves. It was time to start seeing clients.

A Few Hypotheses for the Left Brain

As a human of at least normal intelligence I see, feel or believe:

A. We are not the most advanced beings, nor are we the only beings in this or any other universe.

B. Space ships exist. Many people from many places have seen them and have documented their existence for centuries.

C. Alien contact and abduction are real. The people who have this experience need to be treated with as much care, privacy, unconditional love and non-judgment as a family member, friend, medical doctor, therapist or psychiatrist can muster.

Below are listed a few hypotheses for you to consider while proceeding.

- **You**, The Contactee or Abductee **have a choice** and can stop all contact with lower level beings.
- The contact or abduction experience is one of choice and agreement.
- This experience **seldom, if ever,** occurs

without the consent, conscious or otherwise, of the participant.

- Earth is a free will zone.
- It is against Universal Law for other beings to interfere in this free will zone.
- Some contact/abduction is positive and of a very spiritual nature. Some are not.
- Both positive and negative experiences can occur in a person's life at the same time.
- There are many civilizations in our universe and in universes we never have dreamt existed.
- Some of these civilizations exist in the third dimension, some in the fourth, some in the fifth and so on.
- There are but a few intrusive civilizations. The majority of the civilizations are in alignment with Universal Law.
- A soul may leave the body and go to a place of rest during a time of trauma in early infancy and an alien soul may enter the body to keep it 'on track' until such a time as the original soul is prepared to return to the earth plane.

- Many contactees are being educated by highly evolved beings to aid in the future transition of earth into another dimension.

- Illness, such as some forms of Environmental Disease, Cancer, Chronic Depression, and many more come from the breaking of one's auric field.

- When an abductee's auric field is broken, negative entities and entities caught between worlds enter the aura. Therefore, contacting a person who does entity removal long distance, or for a spiritual check-up, is advisable.

- Aiding the abductee or contactee in regaining physical, mental and emotional control in their life is the first and foremost part of therapy. Then, if the client wishes to regain veiled memory they can do so safely, knowing they now have control of their reality.

- Each person walking his or her path on the earth plane has a right to a joyful, productive and purpose filled life!

- There are people who create negative events and experiences with aliens out of a need to feel wanted by someone or something. This often begins when they first read stories of abductions. (No matter what is done to help the person with the negative experience it will not stop until the emotional lack in his or her life has been healed. Usually through work with the inner child.)

I will not attempt to prove any of the above hypotheses. I am offering something for you to consider, and search for answers within yourself, based upon my experience with clients.

You are going to read very little in the area of case studies. My purpose here is not research, but to share with you what I have learned from my clients.

Well, with these thoughts in mind, are you ready to begin our journey?

Great! Let's begin with a question just for you the abductee, contactee, family member, doctor or therapist.

Have you or your clients experienced any of the following in either a waking or dream state?

- A close approach by a UFO;
- Sudden development of E.S.P.;
- Hear modulated tones in ears;

- Telekinesis abilities;
- Shoulder, knee & back pain;
- Feel earth is not your home;
- Loss of weight for no reason;
- Experience bed shakes;
- Missing time;
- See visions;
- Hear voices;
- See an unexplained light in your bedroom (white or any other color);
- Had a last breath feeling then went immediately to sleep;
- Unexplained scars, rashes or marks on your body;
- Feel your parents are not your real parents;
- Tried to get out of bed but been held down by an unexplained weight on your chest;
- Must sleep with a radio, television or night light on for company;
- Erratic sleep patterns;

- Awake in the morning feeling drugged;
- Waking to electrical appliances not working for no apparent reason.

If so, you may be a contactee or abductee.

Let's Define Some Terms

I have given you a list of experiences to consider. Sometimes I am asked what certain things mean so I feel it would be wise to give you, some definitions.

A Close Approach By A U.F.O.

Let's imagine you are driving down a road during the daytime and you see an unusual looking object in the skies above you. The closer you look you realize it is a shape you have never seen before. It moves very slowly ... then picks up speed; it moves very quickly ... then slows down; it goes straight up or straight down or moves in very unusual patterns. This object or craft could be identified as an Unidentified Flying Object (U.F.O.).

Sometimes craft like these come very close ... close enough to leave a slight sunburn on one side of your body; sometimes they stay a distance away but are still well within range of ones normal eyesight.

Sudden Development of E.S.P. And Seeing Visions

Extra Sensory Perception (E.S.P.) is defined in Webster's New World Dictionary as: adj. apart from, or in addition to, normal sense perception.

A very common E.S.P. experience is when you are walking down the street you might get the feeling someone is following you or looking at you. When you turn around to look you realize this feeling was real ... someone was following you or looking at you. Another one is when you get the feeling you shouldn't go down a certain street or go in a certain direction and sure enough, on the nightly news there was a car accident on that street or freeway.

Vision is defined in Webster's New Dictionary as: n., 1. The power of seeing 2. Something supposedly seen in a dream, trance, etc. 3. a mental image 4. the ability to foresee something as through mental acuteness 5. something or someone of great beauty.

We are currently addressing definitions in which one and five do not apply, but the rest do and are self explanatory.

Hear Modulated Tones In Ears

This one throws a lot of people for a loop. It is also one you need to get **checked by a physician.** If the physician does not find a definite infection he/she can identify, you are most likely being contacted by higher level E.T.'s. Just what you wanted to hear, right?

The pitch seems to usually be of a higher

frequency and occurs sporadically. Many of my clients have experienced this ringing in their ears since early childhood. This does NOT apply to everyone. It can start later in life.

Telekinesis Abilities

A great example of this is walking into your home or office and suddenly light bulbs or fuses start blowing up or your mail never reaching its assigned destination. Believe me, its not always the postal service's fault. Another common example is picking up paper or an object and it seems to just fly out of your hands. Yet another you are picking up a dish and it breaks for no apparent reason.

Shoulder, Knee and Back Pain; Feel Earth is not Your Home; Loss of Weight for No Reason; Feel Parents Are Not Your Real Parents; Awake in Morning Feeling Drugged; Erratic Sleep Patterns

The six feelings or experiences listed above are self explanatory.

Regarding the weight loss, this should be **checked out by a physician before you decide there is no reason for it.

Experience Bed Shakes

This is a very common experience for many people. You know there has been no earthquake, no semi truck passing by and no one wrestling on the bed, yet the bed shakes in the middle of the night. Most people attempt to brush this off as a dream experience.

Missing Time

There are many books written that give individual accounts of missing time, but I have found the scenario described below to be the most common I have come across.

John and Jackie left at seven in the morning for their aunt's house. Aunt Jane's home is only four and one half hours away, if there is heavy traffic. When they arrived it was nearly two o'clock in the afternoon and no one could figure out what took so long for the drive. The traffic was VERY light; there were no car problems or stops along the way.

Hear Voices

There are times in peoples lives when they are being opened up, so to speak, psychically. One of the descriptions heard regarding this 'opening' is the hearing of voices in ones own head. With a few of my clients this has ended up being the explanation of the 'voices'.

But ... what is happening when you are laying in bed, almost asleep, and hear voices in the other room. You get up to check it out and find no one there. The next morning the whole thing seems like a dream, yet you feel a sense of fear. The next time you hear the voices they are in the room you are in. Somehow in both instances, you know you are supposed to be asleep but think you might not be. Just another one of those dreams? I wonder.

See An Unexplained Light In Your Bedroom, Out of Your Window Or Surrounding Your Home

Most of the people I have spoken with, express this experience as occurring in their bedroom and seem to feel it occurred in a dream state. A few have looked out their windows and seen an extremely bright light surrounding their home and retreated, closing the curtain or blinds in total fear of the unknown.

Had A Last Breath Feeling Then Fell Immediately To Sleep

In general, this seems to be as if sudden exhaustion takes over the body, and the person falls into a deep sleep as soon as their head "hits the pillow." I have come across very few who have expressed this experience, consequently I do not feel adequately prepared to address it. However, it has come up a few times so it definitely bears mentioning!

Unexplained Scares, Rashes, Marks or Bruises On The Body

Have you ever gone to bed without a mark on your body and wakened in the morning with a bruise on your leg or a rash in the solar plexus area or somewhere else on your body? Or marks that may resemble an inoculation of some sort on the spine or thigh(s), or some other spot on your body? This is what this experience is referring to.

Tried To Get Out Of Bed But Been Held Down By An Unexplained Weight On Your Chest

Most of my clients have expressed this scenario as a horrifying nightmare. It seems as if fear comes over them, they fight for the voice to scream but can't. The weight on the chest seems to increase the more they fight to get up. They are paralyzed, unable to reach out to their mate for help. Suddenly, the weight is lifted and they wake up in sweats and filled with fear.

Waking To Electrical Appliances Not Working For No Apparent Reason

You go to fix a piece of toast but the toaster won't work, or turn on the television for your favorite morning show but there is no electricity. A few minutes later you give this a try, again, and the appliances work!

The Dream State

Nearly every case of abduction or contact has been in one of two states. The most common is what the client feels is a dream state. In the next few pages we will look at two typical dream-time scenarios recorded by clients in their dream logs.

Dream Scenario - One

Note: The dream scenarios were recorded directly from the client's journal tapes.

First there were these two balls; they were playing in the sky, bouncing, big balls ... they were just beautiful, shimmering balls and they started coming closer and closer and my sense is - oh ... I can't add senses ... and they [the balls] came down closer, and I started saying you cannot take me, it's against Universal Law; you cannot take me, it's against Universal Law. They [the balls] were very persistent and came closer and closer, and I could feel myself going into fear and wanting to give in but I, I started feeling very tired; like I wanted to just go to sleep as I watched the balls. Then I knew what was going on and I finally made myself say one more time, you may not take me, it's against Universal Law, and the balls left.

The scene progressed and I'd gone back inside my apartment or somewhere, or someplace; it was later that same night I think and, ah, ah a little machine had come into the apartment; a little mechanical toy and it was shooting like little darts; actually they [the darts] were more like little staples and I could feel the staple hit my body and said you may not do this it is against Universal Law. I have no contract or agreement with you ... and they kept trying and finally once again I reached this point of wanting to give in and said that if you choose to do this you will pay the karmic debts because it is against Universal Law and the darts were stopped and hit the floor and I was able to pull them [the darts] physically out of my arm, out of my throat and out of my back and the ones that hit the floor were just left there and this thing had left some little darts that had been thrown on the floor and one looked like a little bug that was just ... just there. I could see this [this machine or bug] leave it on the wall and then it [the dart] just dissipated and the next scene was like at a beach, and it was so funny, it was a very handsome man with jet packs on, flying through the air with a dart gun [in his hands] shooting darts or staples and they would just go right through the building, through the thing I was in at the beach. It was like a shop of some kind and these things would come right through the wood or whatever it [the building] was made out of and he was bombarding 'us' and again I was saying it was against Universal Law ... you have no right to do this.

He [the man with jet packs on] was very attractive. Finally, he flew right up to me and pulled a dart [staple] out of my arm and I said if you wish to contact me with information do so telepathically or at a point where we may speak very clearly, but you may not attack or place instrumentation in me every again...or even try.

Then he just softly touched my cheek and flew away. I feel he was telling me I had passed some kind of a test and would be bothered no more by any attempted abduction or implantation of instruments. When I woke up I was completely in sweats and felt my children were going through the same thing but, I know they won't listen to me yet.

Dream Scenario - Two

I ... I am floating out my window. How can this be? I live on the sixth floor of an apartment building. There are two silver balls bouncing in front of me, I feel so tired. Now I am in some sort of a room with no walls that I can see and there are 'people' putting me on a table. I feel scared and ask them not to hurt me. Then the bouncing balls come back and I fall asleep. Somehow I think I wake up ... I feel helpless; no one can hear me ... no one will help me. There is something being placed in my throat and some kind of stinging in my left leg. Now there are owls and wolves floating past me ... where did they come from I asked, but no one answered.

Then I was in an elevator, I think, or the blank room started moving down and then I remember floating back into my apartment. When I woke up I was covered in sweat and my pajamas were sopping wet. I have a sore throat and a mark on my throat like a small incision was made there and there are three marks on my leg ... they look something like what a needle gun makes when one gets an inoculation for measles or something.

The first dream scenario comes from a person who had been working with agreements regarding abduction/contact and had gained the understanding that this sort of thing doesn't happen unless we, as humans, have made an agreement to be experimented with. It is against Universal Law for any other being to interfere with those who live on the planet earth without our consent. This understanding enabled the first client to survive the experience and stop the abduction.

As a side note, the first client is no longer plagued by such dreams and is living a much more balanced and healthy life, and most importantly, this person states that she is in control of her life for the first time since she was a small child.

The second dream sequence is typical of many who live in a large city where there are high-rise apartment buildings. I believe it is important to let you know these abductions/contacts don't just happen on lonely roads or in small country towns. This may be happening to one of your family members or your neighbor right now.

Many people have referred to the earth as a free will zone. If any of the above scenarios seem familiar just remember, you too live in a free will zone and it is **Against Universal Law** to do anything that interferes with one's freedom of choice!

A Family Affair? You're Kidding!!!

Have you ever found your child hiding in a closet, sobbing, "Don't let them take me!"?

Does your child have frightening nightmares about strange beings?

Has your child been told to be home at a certain time and shown up late ... swearing he/she was home on time? [We're not talking about a teenager on a date or out with their friends.] The child couldn't understand why you were angry at them for being late because they left Johnny's house in plenty of time to get home?

Does your child ever have strange tales of someone in their bedroom in the middle of the night and they are so frightened they no longer want to sleep in that particular room?

Does your child tell you of their invisible friends?

Do you allow your child to tell you of their invisible friends or playmates without ridicule, or do you tell them that is impossible, there is no such thing as invisible playmates!?

Let's take a look at the invisible playmate for a moment.

Often, I hear tales from the children of contactees and abductees about invisible friends. They tell me how much company these playmates are when mommy or daddy is at work, or cleaning house or watching a football game or on the phone or ... on and on and on.

One of my first questions in response to this is to ask the child if the invisible friend is their Guardian Angel.

If the child doesn't know I ask him or her to ask their friend to be with them and then simply ask their friend to identify itself. A typical scenario of this sort of discussion with a child under the age of eight would go something like this:

"Will you please ask your friend to be with you for a minute?"

"She's here."

"Please ask your friend if she's your Guardian Angel and ask her to show you a sign that she has come in the name of our Lord Jesus Christ."

"She said yes and her name is ______."

"Did she show you a sign?"

"Oh, yes, she showed me a ____." [often times the child draws the shape of a cross in the air for me rather than using the word cross.]

If any other sign appears it has to do with the

religion the child is raised in. For example, if the child is of the Jewish faith, she is sometimes shown the Star of David or some other symbol that is acceptable to his/her faith, or a bright, white circle with a cross in it.

If the being present is NOT a Guardian Angel it will either show no symbol at all or it may show many other symbols, but these are not appropriate and the being needs to cease contact with this child.

I will add at this time that if the being or spirit does not say it is the child's Guardian Angel, but another spirit, it also needs to be challenged in the same way and must meet the same criteria.

Cases in which those we call E.T.'s. or Aliens, for the lack of a better term, the scenario is virtually the same. This E.T. or Alien will say it is not the Guardian Angel and must be challenged in the same manner. If it does NOT meet the above criteria then the E.T.'s contact with the child MUST be stopped.

The breaking of this contact with a child seven years of age or under is done through the parent. If the child is eight years of age or older, he or she has the right to make the choice to stop this contact. The parent may not make the choice for this child because it interferes with the karmic path of the child.

Briefly reviewing the experiences allows the child to remember and he/she discovers whether they were positive or negative. I explain that they have free will and no longer have to accept this contact, then we begin finding out why, how and when the contact began and stop it. This process is done in a conscious state, not under hypnosis, and the child and parent remain completely aware of the process.

I would like to share a couple of stories about children eight years of age and older. This may give you a better understanding of the responsibility levels of these children.

We will start with Karen. Karen is eleven years old. Three months after Karen's experience her parents brought her to me.

One night she was alone in her bedroom reading. She looked up from her book and saw a figure standing in the doorway.

Karen proceeded to explain to me that she was not frightened at the time because she felt there was no reason to be. Besides, she guessed it was her imagination playing tricks on her.

The next morning she told her mother what had happened, but explained she really couldn't remember if this being came all the way in the bedroom with her. She was unable to remember the full experience, consequently she became deeply fearful and refused to sleep in her bedroom.

I asked Karen if she wanted contact with this being and she responded in the negative so we looked at her agreements for this contact and changed them.

Karen is old enough to decide for herself if she wanted to look at the experience under hypnosis. She felt she did. I asked both of her parents to be present during the session.

Interestingly enough she had absolute recall of this night and several more contacts in rapid succession. The E.T. even introduced himself. None of the contacts were negative in nature.

This E.T. told Karen he was coming to her to protect and teach her and be her friend. He also explained that he had been with her since she was three years old and was sorry she had become frightened.

Karen drew pictures of all of the contacts and found out this E.T. was helping her with her homework, [mathematics], and giving her information regarding the universe. The fear came up because she had recall of one experience but the other contacts had apparently been veiled from her conscious memory.

When the session was over Karen tried to contact the E.T. and it refused to show itself to her. She spoke with her parents and explained that she liked the idea of her friend being around since he wasn't trying to harm her and asked if they would give permission for her to have contact once again. They agreed.

Once again, with the parents present we made a new agreement for Karen to have contact with this being. The agreement made had the following stipulations in it.

1. This being must always show itself and a sign that it worked in the Light of Christ before saying or doing anything;
2. It may never attach itself to her psychic energy system in any way;
3. There is to be **no** experimentation or monitoring of the physical human form;
4. It may not bring any of it's friends without the express consent of both Karen and her parents;
5. There is to be absolutely **no** 'abduction' of Karen in the physical or dream state;
6. There must **never** be any harm to the mental, emotional, physical or spiritual

bodies of Karen. (The E.T. agreed to all the stipulations without hesitation.)

I have been in touch with Karen and her family on a regular basis and all agreements are being met.

A note to this story is that Karen is highly creative in all endeavors and is doing senior high mathematics. She also seems to have an innate knowingness of the universe including planets, stars and civilizations that most have never known.

The second child we will call Bobby. Bobby is eight years old and lives with his mom.

I was called because Bobby's mom felt the house was haunted with spirits ... not so nice ones at that.

There was a strange twist to this instance because Bobby is a very bright child with exceptional psychic abilities.

We found there was definitely alien activity in the house but it was coming from a different source. Bobby was inviting them in alone or with other spirits. Without any discernment!

An interdimensional door was located in his bedroom, one which Bobby opened when he was bored. I asked his mom to see a psychic I knew who is very clear and powerful.

She explained that it was time for Bobby's mom to take control and let him know this was not acceptable. He is now learning not to open this door and the activity in the house has ceased.

If you have children seven years of age or under, you the parent(s), can stop the contact/abductions.

The process is the same as the one you use for yourself; you only have to ask to see the agreements of your child and request them to be updated.

If the contact/abduction is negative you put a stop to them. If it is positive I would suggest you write new agreements for the child as you would for yourself. [Please refer to the exercise section of this book.]

Remember, if one or both of the parents are experiencing contact or abduction the children are most likely having similar experiences. **Contact and/or abduction truly is a family affair!!!**

Walk-Out? Walk-In?

Did you experience an abusive childhood, emotionally or physically? Was there an experience as an infant that may have traumatized you so, that your soul left your body through your body being connection to find some sort of familiar safety? Are you a **Walk-Out?**

Are you from the planet earth? Are you from somewhere in the vast Universe? Do you feel earth is not where you belong? Is it hard for you to comprehend the barbaric ways and beliefs shared by many on this planet? Do you stare deep into the night sky wondering when you will be permitted to leave for home? Are you a **Walk-In?**

Do you experience confusion regarding sexual preference? Is it difficult to feel your own body, in part or in full? Do you wonder why you are here? Do you feel your parents are not your real parents? Are you a **Walk-Out/Walk-In?**

Do you wonder if I am crazy for asking you these questions? Can't say as I blame you! If, out of nowhere, someone started asking me these questions I would wonder the same thing.

Many have heard of the concept of spirit walk-Ins. Ruth Montgomery wrote about this several years ago in her book titled **Strangers Among Us.** You may find it helpful to read this book.

Anyway, we are not going to discuss spirit walk-ins here, but what about the possibility of an alien or extraterrestrial inhabiting and sharing the human vehicle for the purpose of helping the original soul to remain in the third density to complete its mission, so to speak?

Early last year I was given a process to use with my clients. This process helps an abductee/ contactee regain balance in their lives. The process is guided from all sides, spiritual and extraterrestrial.

This process is done in a conscious state, and the client sees and makes decisions based on what occurs for him or her. I facilitate the person getting where they need to be. One might call this process "conscious out of body travel."

What happens is simply that the client remains firmly grounded and is guided through the Universe to a specific vibrational grid. At this level the client calls for their own soul to come forward. If their soul is there it appears; they converse with each other to find out what happened; at what age, and so on.

If the person wishes to have their own soul re-enter the physical body, negotiations are begun. The alien or extraterrestrial soul is usually quite happy to be freed of it's earthly responsibility and leaves within a few moments.

Interestingly enough, there are beings who come to meet the departing soul. They may well be its parents or some other family member(s), to welcome and escort it back to its place of origin.

The client is often given information about the origins of this alien or extraterrestrial soul before it leaves. At this point the true soul of the person re-enters the physical body.

My experience as a healer, made it difficult to use this process at first. Somehow I felt as if I were playing with people's lives and that is not my purpose here. Most of the work I do is guided from a higher source and I had no one, in physical form, I could call to ask for information regarding the use of a process like this.

Consequently, when I presented this concept to my clients I had no hesitation in explaining there was no proof, no study of the effects of the process and, if they wanted to try it, it would be necessary for them to keep a detailed journal of their life perceptions for a period of three months to a year.

In this journal, I have asked them to keep notes on everything from their perceptions regarding family and personal relationships to politics, sexuality and physical health as well as emotional changes.

The clarity and depth of the journal information varies with each person. I would like to share some of their perceptions with you.

The most common and immediate physical reactions are:

- Extreme brightness in eyes.
- Skin color is much healthier.
- A feeling of gaining ten pounds due to being completely present in the physical body.
- Increased circulation.
- Physical movement is easier.
- Physical balance is greater.
- Being able to feel parts of the body they could not feel before due to numbness. Most often amazed at feeling their hands or feet.

The most common and immediate emotional reactions are:

- There seems to be an immediate sense of extreme joy and grief inter-mixed. [The joy seems to come from an inner sense of self. The grief comes from a sense of loss of an old friend.]

- The world seems new. [It's expressed by most as if they were looking at the world for the first time; Colors are brighter, air feels different, and so on.]

The **long term** results of this process are gentle and penetrate deep within the core of one's being. Listed below are some of the **on-going perceptual changes** noted in these journals.

- An ever growing sense of who you are;

- Sexual preference may change. [In all but two cases where the client has been of homosexual persuasion the person involved now feels more comfortable in a heterosexual relationship.];

- Able to cope easier with society as a whole;

- Political views often change;

- A sense of being more in touch with the earth for the very first time;

- Clearing of cellular memory is much easier;
- An on-going spiritual, physical, mental and emotional synthesis.

This process is also used with those who feel they may be a walk-in. Because it is conscious there is no interference from outside the individual and the information remains their own. If a person is a walk-in, they are told so at this time; they may meet their families, find out the reason they choose to be here if it is not remembered, and so on.

This process is not used with every client. There needs to be strong indications that the person is either a walk-in, and in need of knowing this for certain, or exhibiting specific signs that would indicate he/she may be a walk-out. It is never used with anyone due to pure curiosity!

Physical Disease
Entity Attachment
Entity Possession

In 1990 I saw my first client who was suffering from Environmental Illness. Much to my surprise during our initial consultation I found she had marked many of the most common experiences listed on the interview form regarding U.F.O. contact and abduction.

The more we talked, the more I wondered if the symptoms she expressed had anything to do with the alien experiences. I explained my feelings and asked if she would be open to following the usual processes I request of abductees. She agreed.

The process begins with the client writing a letter to an associate who does depossession work.

By the time the depossession was complete she had made a major shift for the better in her health. The majority of the initial working done was repairing her auric field.

In this particular case there was no major depossession needed but that is untrue of the majority of the people I see.

The energy field surrounding our physical body is called the aura. The aura is in place to protect us. When it is broken, illness and entities have the opportunity to enter, not only this field, but the physical body.

Let's briefly explore the causes of 'whacks', or 'tears', in the aura. If a person has undergone any surgery; is given or taking physically or mind altering drugs, (prescription or otherwise, including alcohol); experiencing stress, anger or negative thought forms, the aura is easily broken.

A simple example: if you are in a bar getting bombed, you leave yourself wide open for a discarnate entity that has not gone to the Light, to attach itself to you.

If you are doing drugs you are asking for the same thing. If you are in a hospital and a person passes on after an accident or something, they may attach at that time as well.

What is a discarnate entity? Why doesn't it go to the Light? Why would it attach itself to me? Valid questions! I will attempt to keep the answers as simple as possible.

If a person experiences a traumatic death the soul may not accept that it is no longer in physical form. If this happens, the soul may not accept its opportunity to go to the Light. Consequently, it remains within the fourth dimension.

Not realizing that the soul has passed on, (left the earth plane for the fourth dimension), seems to leave the soul searching out ways to experience life. Hence, it 'attaches' to a person whose aura is broken. A person, in time, may develop this entity's physical illnesses and begin carrying its emotional baggage as well.

If the above concept is confusing to you, you may gain clarity and understanding through reading **The Unquiet Dead,** by Edith Fiore. This is an excellent book indeed!

Are you ready for more? Well, the next concept is one many people will resist: demonic or satanic possession. I am not going to get into this subject in this book except to mention three points.

1. You may be carrying one or more of these beings.

2. You may have brought it in with you from a past life.

3. It may have been placed with you through a dedication to the dark side by some sort of satanic cult, coven or religion.

Since you do not have to have conscious knowledge of any of this it may be a good idea to get a spiritual check-up!

Back to the aura ... after my client's depossession, we cleared many agreements for the abductions and physical experimentation on her body. As she improved she began referring more people with Environmental Illness.

The point I wish to make here is simply that every person she has referred is an abductee. Every person these clients have referred are also abductees ... all have one form or another of Environmental Illness.

So far the major help I have been able to give begins with repairing the aura. This seems to begin a healing process which allows the client to gain more physical strength.

The next step on the physical level is a process I was given, through Higher Guidance, of neurological and genetic reprogramming.

So far some basic results are as follows:

- Lower doses of medications.
- An ability to read again.
- Loss of some food allergies.

- Handwriting improvement.
- Improvement in thought patterns, (able to think clearly, focus or meditate where the person could no longer do this.)
- Able to go out of the home.
- Able to socialize without allergic reactions to perfumes, etc.

Time has passed and all I can do is send notes and/or tapes of the sessions, including the depossessions that are done, to the doctors of these clients. I have no idea if the doctors read the notes or ever listen to the tapes.

It is my hope that one day soon I will meet a doctor who will communicate with me regarding his/her patients ... my clients.

I Agreed To What?!

Note: Self responsibility and courage lead to self healing and self-empowerment. You have a right to a joyful life without physically and emotionally intrusive beings messing with your reality!!!

Yes! You did! You agreed to be contacted, abducted and experimented on. I would venture a guess that ninety-nine percent of the abductions that occur are carried out because of an agreement made between the abductee and the aliens involved. It takes a lot of courage to accept responsibility for such a thing.

OOPS ... Before you slam the book shut let me attempt to explain:

A theologian may argue the exact date but, if I remember correctly, reincarnation was removed from the Bible around the year 555 A.D. during the Council of Nicea. If this is true, and I believe it is, then we live many lifetimes during our evolution.

Some contact/abduction agreements were made centuries upon centuries ago and are honored to this day! Some were made while you were in spirit form. Possibly between lives or before you ever chose to incarnate here!

Hence, whether the contact or abduction scenario is positive or negative, one is responsible for the situation he or she finds themself in.

Consequently, when I began doing regressions with abductees I was not completely surprised to discover agreements had been made by them with alien civilizations, to allow themselves to be used in scientific experiments.

With this thought in mind do you, the reader, understand that one has the right to change or cancel these agreements?

When I bring this up to my clients they are shocked and do not mince words in letting me know I am crazy! Not one of them would ever KNOWINGLY put themselves in this position!

Ah, yes, the key word: KNOWINGLY.

If the original agreement was made in 891 B.C., and reaffirmed during several lifetimes, would a person remember it in this lifetime? I doubt it, so please don't be hard on yourself.

It's definitely difficult to consider one's choice to be a victim in any situation, much less the one under discussion here. It would take many, many regressions to locate one's core agreement in order to begin changing them. Through regression one would have to locate one lifetime after another, changing each agreement as they proceeded.

I have written a few exercises for you to use. One of them, Accessing Your Akashic Records, is designed to aid you in the location and disposition of these agreements.

I can hear the comments now!!!

Nothing has ever worked; I can't do this by myself; I'm not psychic, and so on. I'd like to encourage you. You are your own best psychic! The most valid information you ever receive about yourself comes from your own heart and soul!

I have used this process with many, many people over the last few years, and it works! Each person is in a conscious state and each one gets their own information. I guide them through the process but they are in control! You deserve to have your personal power back and this is one way of doing it!

For more information regarding choice one might want to read Chapter Nine in **The Prism of Lyre; An Exploration of Human Galactic Heritage,** by Lyssa Royal and Keith Priest. This may well give you a much clearer understanding.

Sooo ....

It is not my intent to present myself as an expert in this field. It seems there are no patented answers for anything in life, including personal experience.

Yet I believe it is time for the populace of the planet earth to recognize **we are not alone out there or here, nor are we such an advanced civilization.**

If we were there would be no greed, hence no war; no need for a 'privileged few' to feel they have a right to control the masses and no hunger or poverty. We would live in a state of unconditional love and respect for one another and our planet.

What I have attempted to present are but a few hypotheses that for me have become reality. One person's reality may not be the same for another; it is only a perception and must remain flexible and open to change.

Consequently, we do not have a right to negate another person's experience just because we have not knowingly lived it in our life.

Sooo ... ...

The clients I have seen regarding contact and abduction have some points in common in regard to their experience. A few of these similarities were listed in the first section of this book.

It's important to note that all abductees/ contactees do not mark all of the experiences listed, but experiencing even one suggests he or she may need to look at the possibility of this being a reality in their life.

Not all contact or abduction is traumatic or negative, but pain in ones life is.

Pain, whether it is physical, mental emotional, or a combination thereof, makes if difficult to remain grounded and centered in the physical body and the pain needs to be healed.

It is impossible for one person to carry the knowledge to help in all areas of healing. Because of this I have made it a priority to bring together alternative healers to refer my clients to. Healers that remain non-judgmental; healers that use abilities not necessarily accepted by traditional medical and psychiatric communities.

This includes people whose expertise is in Body Work, (massage therapy), Herbology, Acupuncture or Acupressure, Vibrational (sound) Healing, Aromatherapy, Intuitive Arts, Hands on Healing, Reiki Energy Work, Eastern Meditation Arts, and so on.

It is my understanding that a group exists on the east coast called Treat. This group of doctors and psychiatrists have been brought together by an abductee named Budd Hopkins, and it has been doing some remarkable work in behalf of abductees and contactees. Unfortunately, to the best of my knowledge, such a group does not exist on the west coast.

One point I wish to make clear regarding the contact and abduction experience is that it is, ninety-nine percent of the time, a matter of choice; agreement if you will. This may be the only truth I feel is not open to flexibility.

A second point is regarding childhood trauma. Getting to know, love and work with ones inner child is critical in healing one's life. The majority of these experiences begin in childhood during a time of stress, trauma or loneliness.

The only other truth is that abduction and contact seems to be so common in the United States, that people quite often do not realize they have had the experience. It is written off as a bad nightmare or a moment of insanity.

A great question to ask of doctors in a psychiatric hospital is: **How many of your patients tell you they have been abducted by U.F.O.'s?**

You may be extremely surprised at their response!

Should doctors be treating these patients with different methodologies? Quite probably.

Most importantly, we as people, need to get out of the doom and gloom mode regarding contact and abduction.

Once the contactee or abductee regains control of their lives, stops playing into the hands of lower level intelligence, and opens up to the higher level of contact they may receive information that aids in their personal lives and can help bring balance to this planet.

It is my hope that in the very near future, members of the medical and psychiatric communities will be open enough to delve more deeply into the area of U.F.O.'s; that they will not be fearful of ridicule from their peers; open up, consciously, to their intuitive abilities and recognize how important it is to work together to heal the whole being of a person.

What Defines An Alternative Therapy

Note: It is a doctors', psychiatrists' or therapists' responsibility to look at and use, all available technologies that are not physically harmful, (regardless of personal belief systems), including Alternative Therapies to aid a patient or client in a speedy recovery!

I believe the preceding statement to be an undeniable truth!

I also realize there is little, if any, current scientific proof that exists in western society, that is acceptable to many in the medical or psychiatric communities, regarding the use of Alternative Therapies, but maybe these people need to start researching tools available from the past.

Okay, so what is an alternative therapy and why is there a need for the traditional psychiatric and medical communities, and the alternative therapies communities to work together?

A general rule of thumb is that an alternative form of therapy does not entail the use of chemically related drugs. It is the holistic approach to the body, mind and spirit of the person.

An alternative therapy can productively be used in conjunction with traditional medical and psychological treatment, or simply for personal growth.

Where the physical body is concerned we have massage therapy, hands on healing, Reiki, acupressure, acupuncture, and chiropractics to name but a few alternative therapies available to us today.

Where the subconscious mind is concerned, that is, our mental, emotional bodies, the traditional form of treatment has been psychotherapy. Today we have a new tool to add, Alchemical Hypnotherapy.

Science as we know it today had to have a root form, a beginning. Somehow, over the centuries man has forgotten his roots, his ties to the earth but, must now regain this lost knowledge and incorporate it into current technologies in order to heal. Just in case no one is looking, there are illnesses doctors can't find a cure for. Maybe we have polluted the physical form with to many foreign chemicals.

It is my personal belief that most forms of therapy have their place and need to be used in harmony with one another. The ancient traditions of Hands on Healing and Shamanism are the root of modern medical and psychiatric treatment.

Once we bring the Medical, Psychiatric and Alternative Therapies Communities together, we will be in a position to heal and bring into harmony, all our parts: the physical, mental, emotional and spiritual bodies.

Few remember the following fact:

At the turn of this century Homeopathy was the accepted form of medical treatment, surgeons were considered craftsmen ... it wasn't until the advent of sulfa drugs and antibiotics, (chemically related drugs), **that alopathic medicine came into being.**

Exercises

When you begin to use the following exercises there can be a lot of resistance. The resistance is okay ... honor the resistance and let it go. Most of us have 'hired' resistance as a sub-personality early in childhood to protect us and, believe it or not, Mr./Ms. Resistance has done a great job!

In time, you may want to locate an Alchemical Hypnotherapist in your area to help you in reassigning Resistance to a new position.

It is not necessary to be visual, (see pictures in your mind), to use the exercises on the following pages. Simply use your imagination, (daydream, if you will), and know you are doing them correctly. Mentally telling yourself to do the steps works just fine.

Remember, you are helping yourself and there is no performance being recorded here, so just do them to the best of your ability. With daily practice they will become easier and easier.

The first exercises are to teach you to connect yourself with the place you live, the planet earth. It is important to know and feel these simple exercises in your body. Like it or not you live here, on the planet earth, and if your energy is running around and not with you, in your body, you are wasting it.

Grounding brings your energy back to your physical being and allows you to be centered. This also gives your body a way to release emotions into the earth for transmutation, allowing for positive change within yourself.

Learning the grounding is the single most important gift you can give yourself; with this gift of the Self, you are setting the stage for the successful use of the following exercises.

Exercise #1
Breathing

Our breath is the gift of life. Learning to use the breath to help our body and mind function as we need them to is very important. This first exercise is the first step to empowering yourself and your body.

Breathing, in the way described below, allows you to release anxiety, tension and fear, as well as the wide range of emotions we carry, and hold, within the cellular structure of the body.

1. Ask for Archangel Michael to be with you and protect you during the exercise. [**Please do this no matter what your religious affiliation.**]

2. If fear is present, honor and acknowledge it with the following affirmation:

Fear is present - I release all fear to make room for beneficial change in my life.

3. BREATHE

Begin breathing in through your nose and out through your mouth. Slowly - count to four as you breathe in; hold your breath while you count slowly to four; exhale slowly through your mouth to the count of four.

Follow this breathing pattern for a few minutes, three or four if you can, each time you begin to do any of the exercises listed in this book.

Practice this technique when you are driving a car, watching television, walking down a street, preparing for sleep, or whatever, until you grow accustomed to it.

Exercise #2
Grounding

Follow steps **1, 2 & 3** in the previous exercise.

4.Imagine a golden ball of energy at the base of your spine. (For women imagine this ball of energy at the base of your spine in between your ovaries.) Make it about two to four inches round and allow it to spin for a moment.

5.Allow this ball to send a cord of energy from your body, to the center of the earth. Make this cord at least two inches thick (five and one half centimeters).

6.Imagine yourself connecting the cord to a personalized nameplate. The nameplate says your name on one line and just below your name it says grounded in the present moment. Connect your cord firmly to the nameplate!

Remember To Breathe

7. In the arch of each foot you have a foot chakra, an energy point in the body. Ask these two energy points to open up one hundred percent.

8. Now ask for the healing energies of the Earth to flow in these chakra points and up your leg channels to the ball of energy at the base of your spine.

- If you feel blocks, [pain in one or both of your legs], or if the energy flows slowly, simply give your body permission to release the blocks. 'See' the blocks(s) moving up the leg(s) to the ball of energy at the base of the spine and down to the earth.

9. Ask your auric field to come within eighteen (18) inches [fifty five and three quarters of a centimeter], of your body. (Imagine the aura is doing this. When you ask for this, it happens. **SIMPLY ASK!**)

10. Now, imagine an outline of medium blue color surrounding your aura.

Exercise #3
Accessing Your Akashic Records

NOTE: It is important to remember that there may be millions of agreements for lower level contact. Removing the core agreements is the beginning. Using this process several times will enable you to clear them all. Once is usually not enough. One must always remember to work with these beings with caring and manners. **Be polite, treat them as you wish them to treat you!!!**

Breathe and Ground Yourself by following steps **1 through 10** of the previous exercises.

TRUST YOURSELF and whatever information you are given.

11. Simply ask for the Keeper of the Akashic Records to be with you. Ask him to stand outside of your aura.

12. Challenge the being that comes before you by asking it to show you a sign that it works in either the Light of Christ or the Light of the one known as Sananda (Sananda is the name used by Christ in other dimensions or in the Universe.).

***** Remember *****

a. You **DO NOT** have to be visual. Your grounding will help you feel the presence of 'someone'.

b.The 'sign' you are requesting must feel right for you. It is usually a Cross, a Star of David, a Cross within a Circle or whatever you are comfortable with according to your belief system.

c.This sign is not necessarily visual. You may actually feel energy in your body that seems to flow in the appropriate shape or pattern.

13. If no sign appears at first don't get discouraged. You may well be tested. Other beings that do not want you to stop contact may appear before you in an attempt to 'trick' you. SIMPLY tell the being(s) to get out of your space. If it/they do not leave immediately ask Archangel Michael to remove them. Keep requesting the Record Keepers presence until you are given an appropriate sign.

14. Explain to the Record Keeper that you wish to see all agreements, past life or present life, that you have for alien contact and/or abduction.

15. Ask him if you have learned your lessons from this. If his reply is positive, ask him to please store the lessons and ask him if it is appropriate for all such agreements to be destroyed. If he agrees it is appropriate, ask him to do so immediately.

It is important to **remember** that **NOT ALL CONTACT** is **NEGATIVE** or **INTRUSIVE!** One of the greatest gifts you can give yourself, other than empowerment of the Self, is telepathic contact with the Higher Level Beings of the many Universes and Dimensions, [fifth dimension and above].

There are many civilizations who are willing to aid earthlings in the transformation of this planet to a more peaceful level of existence.

If you wish to open contact with the positive beings simply use the above exercise/process with the Record Keeper and ask him to make up a new agreement between you and these beings.

Such an agreement must be very specific. It may read something like this:

- I agree to telepathic contact with ONLY beings who work in the Light of Christ.
- I agree to enter a 'ship' or a light craft for a period of no more than ten minutes earth time per year.
- All contact must be non-intrusive and may cause no harm to the human vehicle.
- All contact must be for the good of all and mentally, emotionally, spiritually and physically positive.

- I agree to all of this so long as I am permitted conscious memory of such contact [Often these beings will request you to amend or agree to a veil over some of the information if it is felt one's mind is not ready to handle it yet: i.e. a future event that you may be accomplishing but in present time you would feel inadequate.]

The choice is yours. You may not wish to have this contact for a period of time. Know that your wishes will be honored by the Extraterresterial Being(s). These beings follow Universal Law and do not interfere in a free will zone. Their presence must be requested.

One advantage to having contact with E.T.'s. is to request their intervention if the lower level ones, the aliens, keep pestering you.

Aliens, (and humans as well), of a lower consciousness do not want to loose control or power over others. Consequently, they may return a few times to see if intimidation will keep you in their grasp.

I realize all this may sound overly simple. Well, it truly is, it's simplicity itself. Accepting responsibility for your life situation is never easy but once you do, many positive changes occur. These higher level beings are available to help you in any way they can!

Alternative Therapies Resource Network

The people listed below work synergistically toward the healing of others. I am honored to be able to refer clients to them.

Ed Spena, C.M.T.

Ed is a graduate of the McKinnon School of Professional Massage and Bodywork in Oakland, California. He is a powerful hands-on, Christforce healer. His integrated approach to healing includes the following bodywork methods; Shiatzu, Swedish/Esalen, Jin Shin Do, Chinese Message Therapy as well as Second Degree Reiki. As a part of his healing and bodywork practice, Ed teaches Tai Ch'uan, an ancient and timely art of meditation movement.

Mark W. Frost, M.S.C.M., L.Ac.

Mark was classically trained in Acupuncture and Traditional Chinese Herbal Medicine at the American College of Traditional Chinese Medicine, in San Francisco, California. Through the use of Acupuncture, Traditional Herbal Medicine, Corrective Bodywork, and Biomagnetic and Energetic Rebalancing he aids each client integrate healing on all levels: Physical, Mental, Emotional and Spiritual.

Linda Garcia

Linda is a trained psychic who uses her clairvoyance, (clear sight or the ability to see energy), and other psychic gifts as tools for self awareness and change. She uses the psychic information and spirit guides to help with energy healings that go to the root cause of problems to affect life changes. Ms. Garcia is a U.F.O. contactee and can help other contactees/ abductees to gain awareness and control over their lives.

Eva Fried, C.M.T.

Eva is a graduate of the McKinnon School of Professional Massage and Bodywork in Oakland, California. She is an intuitive body worker whose work focuses on release of cellular memory and energy blocks within the physical body and auric field with added expertise in working with impact injury cases. Ms. Fried is an abductee/contactee who aids others in the release of negative cellular memory.

Irene Neumark, L.Ac.

Irene is a Licensed Practitioner of Traditional Chinese Medicine and a Holistic Health Practitioner and Educator. She employes a wide scope of modalities such as: Shiatsu and Breema Body Work; Acupuncture; Herbology; Nutritional Consultation; Vibrational Remedies;

Homeopathy; Aroma Therapy and Therapeutic Baths to facilitate the release of environmentally accrued toxins, and negative emotional crystallizations.

Luis-Antonio Romero

Luis, through the integration of his artistic talents and intuitive abilities aids you to remember who you are and why you are here. His unusual water color techniques create vivid portrayals of your past, present and future lives in conceptual and symbolic illustration. Symbols and colors are designed to activate and integrate your memory of the knowledge of the Ancients with the energies of the present.

Kevin Cates

Kevin, a powerful Christforce healer and musician, combines his many talents in the area of Vibrational Healing. Through his extraordinary use of the Didjeridu, (an ancient Aboriginal instrument), Tibetan Tingsha Bells, Singing Bowls, chanting and harmonic overtoning, he creates vibrational and auditory balancing and release of blocks in the physical, mental, emotional and spiritual bodies. Kevin also leads Earth Medicine Wheel Rituals to aid clients in better connecting with the Earth as well as providing a healing support group.